MW01141483

Grasslands

by Violet Findley

ISBN 978-0-545-24808-2

Photographs © 2010: cover: Peter Arnold, Inc./C. Wermter/Peter Arnold, Inc.; back cover top: Nature Picture Library Ltd./ Shattil & Rozinski; back cover bottom: iStockphoto/Zorani; page 1: iStockphoto/YinYang; page 2: iStockphoto/Bartosz Hadyniak; page 3 main: Alamy Images/Alan & Sandy Carey/Peter Arnold, Inc.; page 3 inset: iStockphoto/Wojtek Kryczka; page 4: Alamy Images/Antje Schulte; page 5: iStockphoto/Stephen Goodwin; page 6: Corbis Images/Jeremy Woodhouse; page 7 main: Photo Researchers, NY/ Jim Zipp; page 7 inset: iStockphoto/Ryerson Clark; page 8: iStockphoto/Zorani; page 9: Corbis Images/Joe McDonald; page 10: Nature Picture Library Ltd./Jose Luis Gomez de Francisco; page 11 main: Getty Images/ National Geographic; page 11 inset: iStockphoto/Erik Lam; page 12 main: iStockphoto/Larisa Lofitskaya; page 12 inset: iStockphoto/Eric Delmar; page 13: Minden Pictures/Yva Momatiuk & John Eastcott; page 14: iStockphoto/David Chadwick; page 15: ShutterStock, Inc./Patrick McCall; page 16: iStockphoto/Nico Smit.

Photo research by Jenna Addesso; Design by Holly Grundon

12 11 10 9 8 7 6   14 15/0

Printed in the U.S.A.   40

First printing, November 2010

SCHOLASTIC INC.

NEW YORK • TORONTO • LONDON • AUCKLAND
SYDNEY • MEXICO CITY • NEW DELHI • HONG KONG

Grasslands are flat and windy places with few trees.

Grasslands are home to a lot of **animal**s. Take a look!

jackrabbit

Fast Fact

Jackrabbits eat **grass**, clover, and dandelions.

What **animal** is jumping **through** the **green grass**? It's a rabbit!

field mouse

Fast Fact

Mice are mammals just like people.

What **animal** is scurrying **through** the **green grass**? It's a mouse!

garter snake

Fast Fact

Garter snakes eat mice, lizards, and insects.

What **animal** is crawling **through** the **green grass**? It's a snake!

prairie chicken

Fast Fact

Prairie chickens can fly up to 50 miles per hour.

What **animal** is strutting **through** the **green grass**? It's a bird!

bison

Fast Fact

Bison live in groups called herds.

What **animal** is roaming **through** the **green grass**? It's a bison!

monarch butterfly

Fast Fact

Butterflies drink the sweet nectar of flowers.

What **animal** is fluttering **through** the **green grass**? It's a butterfly!

screech owl

Fast Fact

Owls hunt for food at night.

What **animal** is flapping **through** the **green grass**? It's an owl!

toad

Fast Fact

Unlike frogs, toads have bumpy skin.

What **animal** is hopping **through** the **green grass**? It's a toad!

coyote

Fast Fact

Coyotes are related to dogs.

What **animal** is slinking **through** the **green grass**? It's a coyote!

ladybug

Fast Fact

Ladybugs can be yellow, black, or orange.

What **animal** is creeping **through** the **green grass**? It's a ladybug!

Fast Fact

Prairie dogs live underground in big groups.

You never know what you will see moving **through** the **green grass**!

Sight Word Review

Point to each sight word. Then read it aloud.

animal

green

grass

through

grass

animal

green

through

Sight Word Fill-ins

Use one sight word from the box to finish each sentence.

animal	**grass**
green	**through**

1 Can you name an _____ that lives in grasslands?

2 Wow, that grass is very _____!

3 A little mouse can hide in the _____.

4 Snakes slither _____ the grass.

All About Grasslands

Ask a grown-up to read this with you.

A grassland is a habitat that has short plants such as grass and bushes. But it has very few trees. One-fourth of the Earth's land is grassland.

mongoose

Grasslands in America are called prairies. Grasslands are called by different names around the world. Grasslands in Africa are called savannahs. Grasslands are also called pampas in South America, rangelands in Australia, and steppes in Asia. Whatever the name, grasslands are the perfect home for many animals.

Grass provides hiding places for creatures such as lizards and mice. Other animals such as meerkats and prairie dogs build burrows in the grass. Because so many animals live in grasslands, predators such as cheetahs and mongooses make their homes here, too. It's an easy place to find prey.

There are also many animals such as bison and llamas that eat the grass. You could say they are eating their own habitat. That's okay—grass grows back quickly, which is another reason that grasslands are great.